Name ______________________ Class ______________ Date ____________

Skills Worksheet

Directed Reading

AF685703

Lesson: Human Reproduction

1. What is a sex cell?

____

____

THE MALE REPRODUCTIVE SYSTEM

2. What is the main function of the male reproductive system?

____

____

3. The sex cells made by males are called ____________________.

SPERM PRODUCTION

______ **4.** Which organ makes sperm and testosterone?

- **a.** bladder
- **b.** penis
- **c.** testis
- **d.** scrotum

CARING FOR THE MALE BODY

5. Explain how males can practice good hygiene.

____

____

____

THE FEMALE REPRODUCTIVE SYSTEM

6. What are the two main functions of the female reproductive system?

____

____

MENSTRUATION

7. What is menstruation?

____

____

CARING FOR THE FEMALE BODY

8. How can STDs be prevented?

____

____

Lesson: Before You Were Born

9. The process by which one sperm and one egg come together to form a new human cell is called ______________________.

10. After fertilization, the developing human is called a ______________________.

CARING FOR THE FETUS

11. Explain how a pregnant woman can help protect the health of the fetus she carries.

____

____

____

Lesson: Infancy and Childhood

INFANCY

12. What is infancy?

____

____

____

CHILDHOOD DEVELOPMENT

______**13.** The stage of development after infancy is called

a. childhood.
b. prenatal.
c. adulthood.
d. adolescence.

14. Describe the three stages of childhood.

____

____

____

____

____

Lesson: Adolescence

15. Define *adolescence.*

____

____

____

WHAT CONTROLS GROWTH?

16. A ______________________ is a chemical that is made in one part of the body and causes a change in a different part of the body.

17. Hormones that affect your reproductive system are called

______________________.

GROWTH SPURTS

______**18.** The stage of development when the reproductive system becomes mature is called

a. infancy.
b. childhood.
c. puberty.
d. adulthood.

PHYSICAL CHANGES

19. List three physical changes that usually occur during puberty.

____

____

____

SOCIAL CHANGES

20. Explain what social changes occur during adolescence.

____

____

____

____

____

MENTAL AND EMOTIONAL CHANGES

21. Who can help you deal with the changes of adolescence?

____

____

____

Lesson: Adulthood, Aging, and Death

______**22.** The period of life that follows adolescence is called

- **a.** infancy.
- **b.** childhood.
- **c.** puberty.
- **d.** adulthood.

LIFE AS AN ADULT

23. What are two responsibilities of adults?

____

____

____

AGING

24. List four factors that affect aging.

____

____

____

____

CONDITIONS ASSOCIATED WITH AGING

25. List three health conditions associated with aging.

____

____

____

DEATH AND GRIEF

26. The end of life, when all necessary life functions stop, is called

______________________.

27. What is grief?

____

____

____

DEALING WITH GRIEF

28. List four healthy ways to deal with grief.

____

____

____

____

____

____

Name ______________________ Class ______________ Date ______________

Skills Worksheet

Concept Mapping

Lesson: Human Reproduction

Use the following terms to complete the concept map below: *male reproductive system, sperm, sex cells, female reproductive system, testes, eggs,* and *ovaries*.

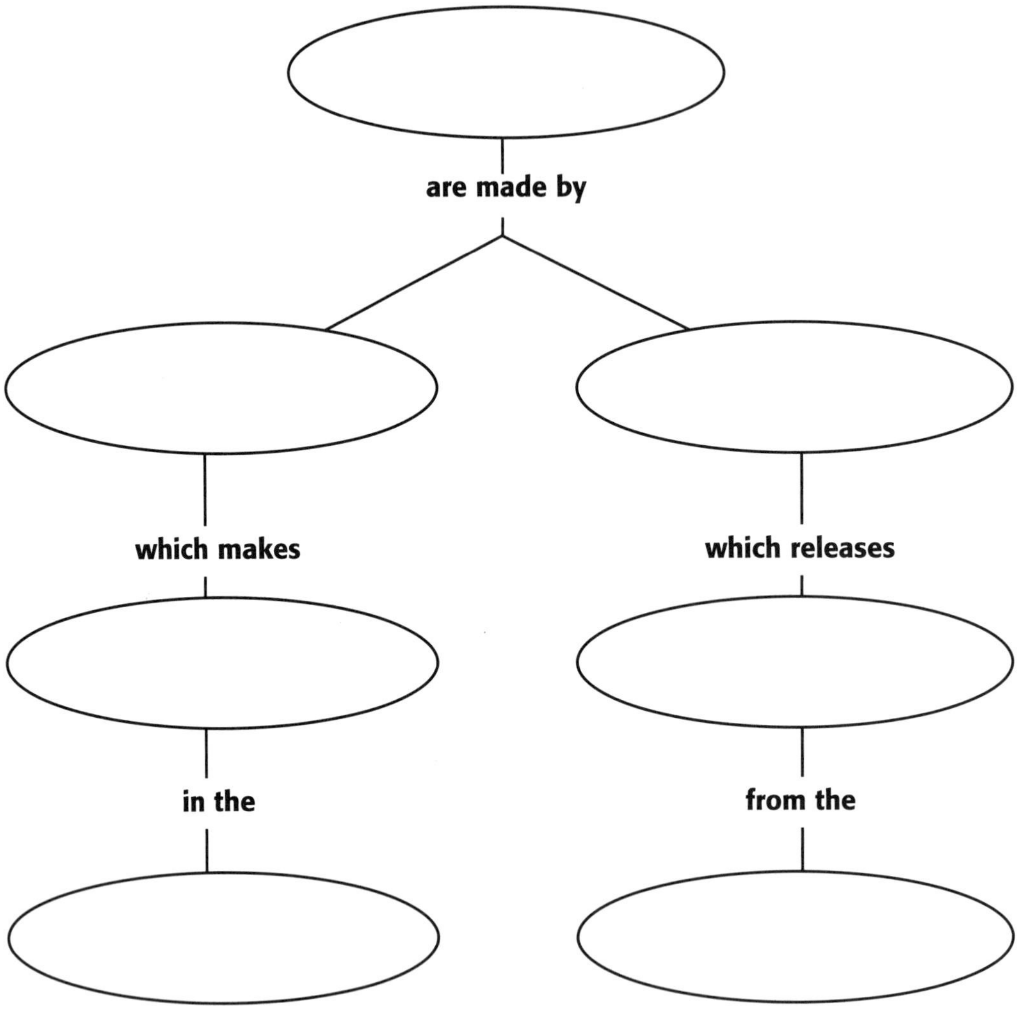

Name ______________________________ Class ______________ Date ____________

Skills Worksheet

Concept Mapping

Lesson: Adulthood, Aging, and Death

Use the following terms to create a concept map in the space below: *adulthood, death, grief, aging, heart disease, osteoporosis,* and *Alzheimer's disease.*

Name ______________________ Class ______________ Date ____________

Skills Worksheet

Concept Review

Lesson: Human Reproduction

______ **1.** The sex cells made by males are called

a. testes.
b. eggs.
c. sperm.
d. ovaries.

______ **2.** The sex cells made by females are called

a. testes.
b. eggs.
c. sperm.
d. ovaries.

3. The ______________________ make sperm and testosterone.

4. The ______________________ are the organs that make eggs and the hormones estrogen and progesterone.

5. The monthly breakdown and shedding of the lining of the uterus is called ______________________.

6. List five ways to protect your reproductive health.

____

____

____

____

____

Lesson: Before You Were Born

7. Describe the process of fertilization.

____

____

____

8. The developing human inside the uterus is called a ______________________.

9. At what point during the development of a fetus are the organs fully functional?

____

____

10. Describe three factors that affect the health of both the mother and fetus.

____

____

____

Lesson: Infancy and Childhood

______ **11.** The time between birth and 1 year of age is called

a. infancy.
b. childhood.
c. adolescence.
d. adulthood.

12. List the three stages of childhood development.

____

____

____

13. What are the characteristics of middle childhood?

____

____

____

Lesson: Adolescence

Match each word in the right column with the definition in the left column. Write the letter in the space provided.

______ **14.** a chemical made in one part of the body that causes a change in a different part of the body

______ **15.** the stage of development when the reproductive system becomes mature

______ **16.** the stage of development in which humans grow from childhood to adulthood

a. adolescence
b. hormone
c. puberty

17. What effects do sex hormones have on the body?

18. Describe three physical changes that happen to boys during adolescence.

19. Describe three physical changes that happen to girls during adolescence.

20. List four mental and emotional changes that occur during adolescence.

Lesson: Adulthood, Aging, and Death

21. Describe adulthood.

Concept Review *continued*

22. Explain how technology has affected life expectancy in the United States.

____

____

____

____

____

____

____

23. List the stages of grief.

____

____

____

____

____

____

____

Name ______________________________ Class _______________ Date _____________

Skills Worksheet

Refusal Skills

Lesson: Human Reproduction

Imagine that you are the advice columnist for a parenting magazine. Read the letter below and use your knowledge of refusal skills to write a response to be published in the magazine.

Dear Know-It-All,

My husband and I have been married for several years. We recently decided to have a child. I just discovered that I am pregnant. My husband wants to have a party where we will be serving alcohol. Many of our friends will expect me to drink with them. What should I do?

Sincerely,

Worried

Name ______________________ Class ______________ Date ____________

Skills Worksheet

Refusal Skills

Lesson: Adolescence

Describe how you would use the following refusal skills to respond to the following scenario. Remember to be clear and choose your words carefully. Describe your body language as well as your words.

Your friend Michael just started smoking. One day you are at his house, and he offers you a cigarette. You are not interested in smoking, but you want to feel cool. He tells you that all the popular people at school smoke.

1. **Say no.** How would you tell Michael no?

____

____

2. **Offer an alternative.** What could you do instead of smoking?

____

____

3. **Stand your ground.** What would you do if Michael kept pressuring you to smoke?

____

____

4. **Walk away.** Describe how you would get out of the situation.

____

____

5. **Plan ahead.** What could you do to avoid this situation?

____

____

____

6. **Have a support system.** Who will stand by you when you are in this situation? How can you use these people as support when dealing with this situation?

____

____

____

Name ______________________ Class ______________ Date ____________

Skills Worksheet

Decision-Making Skills

Lesson: Human Reproduction

Read the following situation. Then, follow the steps below to decide what you would do in this situation.

You are experiencing severe pain in your lower abdomen. You try to ignore it for a while, but it just won't go away. You know you should see a doctor, but you don't like going to the doctor because it's embarrassing. However, the pain is getting worse every day.

1. **Identify the problem.** What decision do you have to make?

____

____

2. **Consider your values.** What is important to you?

____

____

3. **List the options.** What possible actions could you take?

____

____

4. **Weigh the consequences.** List the pros and cons of each option.

____

____

____

5. **Decide and act.** Describe what you will do. Explain your decision.

____

____

____

6. **Evaluate your choice.** How do you feel about the action you took? Did you make a good decision? Would you take a different action if faced with the same scenario again?

____

____

____

Name ______________________________ Class ______________ Date ____________

Skills Worksheet

Decision-Making Skills

Lesson: Adolescence

Read the following story and then write an ending to the story in the space below. Be sure to include some decision-making skills in your ending to the story.

I was invited to a birthday party at my friend Anita's house. We were having a lot of fun. Anita's parents went upstairs. Anita walked over to the refrigerator and took out some beer. She started passing the beer out to all of the guests, who were excited to try it. Then, Anita handed one to me.

Name ______________________ Class ______________ Date ____________

Activity

Cross-Disciplinary: Art

Lesson: Before You Were Born

Make a computer-generated poster that shows each stage of fetal development. You should have arrows that connect each stage. Describe what happens during each stage of development. You can use the space below to sketch some ideas. Then, use a computer graphics program to create your poster.

Name ______________________ Class ______________ Date ____________

Activity

Cross-Disciplinary: Language Arts

Lesson: Infancy and Childhood

Think about your childhood. Select a significant event in your childhood, and write a personal narrative that describes the event. Tell what happened and explain why it was an important event in your childhood.

Name ______________________ Class ______________ Date ______________

Assessment

Quiz

Lesson: Human Reproduction

Match the definitions with the correct term. Write the letter in the space provided.

______ **1.** female organs that release eggs

______ **2.** parent cell that can join with another cell to create a new cell

______ **3.** sex cells made by males

______ **4.** female sex cell

______ **5.** organs that make sperm and testosterone

______ **6.** organ that holds the fetus during pregnancy

______ **7.** fungal infection of the groin skin

______ **8.** monthly breakdown and shedding of the lining of the uterus

a. sperm
b. testes
c. egg
d. ovaries
e. menstruation
f. sex cell
g. jock itch
h. uterus

Name ______________________________ Class ______________ Date ____________

Assessment

Quiz

Lesson: Before You Were Born

Write the letter of the correct answer in the space provided.

______ **1.** The process in which one sperm and one egg come together to form a new human cell is called
- **a.** reproduction.
- **b.** incubation.
- **c.** fertilization.
- **d.** menstruation.

______ **2.** At which stage in fetal development does the heart start to beat?
- **a.** 6 weeks
- **b.** 12 weeks
- **c.** 13 weeks
- **d.** 20 weeks

______ **3.** At which stage in fetal development does a fetus react to sound?
- **a.** 6 weeks
- **b.** 12 weeks
- **c.** 13 weeks
- **d.** 20 weeks

______ **4.** At which stage in fetal development are arms and legs formed?
- **a.** 6 weeks
- **b.** 12 weeks
- **c.** 13 weeks
- **d.** 20 weeks

______ **5.** At which stage in fetal development can a fetus make a fist?
- **a.** 6 weeks
- **b.** 12 weeks
- **c.** 13 weeks
- **d.** 20 weeks

Name ______________________ Class ______________ Date ____________

Assessment

Quiz

Lesson: Infancy and Childhood

Write the letter of the correct answer in the space provided.

______ **1.** During which stage of life do humans learn to read and write and connect ideas?
a. infancy
b. early childhood
c. middle childhood
d. late childhood

______ **2.** During which stage of life do humans learn to crawl?
a. infancy
b. early childhood
c. middle childhood
d. late childhood

______ **3.** During which stage of life do humans learn to walk and run?
a. infancy
b. early childhood
c. middle childhood
d. late childhood

______ **4.** During which stage of life do humans learn to dress and undress themselves?
a. infancy
b. early childhood
c. middle childhood
d. late childhood

______ **5.** Which of the following is NOT a characteristic of infancy?
a. rapid increase in height
b. short sentences
c. crawling
d. rapid increase in weight

Name ______________________ Class ______________ Date ____________

Assessment

Quiz

Lesson: Adolescence

Write the letter of the correct answer in the space provided.

______ **1.** During which stage of human development do humans grow from childhood to adulthood?
a. infancy
b. late childhood
c. adolescence
d. None of the above

______ **2.** A chemical that is made in one part of the body and causes a change in a different part of the body is called a(n)
a. hormone.
b. growth spurt.
c. ovary.
d. gland.

______ **3.** Hormones that affect your reproductive system are called
a. ovaries.
b. adrenal glands.
c. testes.
d. sex hormones.

______ **4.** During which stage of development does the reproductive system become mature?
a. infancy
b. childhood
c. puberty
d. adulthood

______ **5.** Which of the following is a characteristic of adolescence?
a. growth spurts
b. acne
c. hair growth on the body
d. All of the above

Name ______________________ Class ______________ Date ______________

Assessment

Quiz

Lesson: Adulthood, Aging, and Death

Match the definitions with the correct term. Write the letter in the space provided.

______ **1.** a feeling of deep sadness about a loss

______ **2.** change that happens as adults get older

______ **3.** the period of life that follows adolescence

______ **4.** a condition that leads to memory loss and thinking problems

______ **5.** expected age of death

______ **6.** a condition in which bones lose a lot of their mass and strength

______ **7.** when all necessary life functions stop

a. adulthood
b. death
c. grief
d. aging
e. osteoporosis
f. Alzheimer's disease
g. life expectancy

Name ______________________ Class ______________ Date ____________

Assessment

Chapter Test

Growth and Development

USING VOCABULARY

Use the terms from the following list to complete each sentence below. A term may be used only once. Some terms will not be used.

sperm	ovary	menstruation	fetus
fertilization	infancy	adolescence	hormone
death	grief	adulthood	

1. A feeling of deep sadness about a loss is called ______________________.

2. During ______________________, one sperm and one egg come together to form a new human cell.

3. The breakdown and shedding of the uterine lining is called

______________________.

4. The sex cells made by males are called ______________________.

5. A developing human inside the uterus is called a(n) ______________________.

6. The stage of human development between childhood and adulthood is called

______________________.

UNDERSTANDING CONCEPTS

Write the letter of the correct answer in the space provided.

______ **7.** Which of the following is NOT a part of the male reproductive system?
- **a.** vas deferens
- **b.** ovary
- **c.** penis
- **d.** testes

______ **8.** Which of the following is part of the female reproductive system?
- **a.** uterus
- **b.** testes
- **c.** vas deferens
- **d.** scrotum

______ **9**. Which organ holds the fetus during pregnancy?
a. vagina
b. ovary
c. uterus
d. scrotum

______ **10**. Which of the following is NOT an example of protecting your reproductive health?
a. bathing every day
b. visiting a doctor regularly
c. abstaining from sex
d. wearing tight clothing

______ **11**. At what point during fetal development does the fetus begin to move?
a. 6 weeks
b. 12 weeks
c. 14 weeks
d. 16 weeks

______ **12**. About how long after fertilization is a baby born?
a. 20 weeks
b. 30 weeks
c. 40 weeks
d. 50 weeks

______ **13**. During which stage of development do children learn to walk?
a. infancy
b. early childhood
c. middle childhood
d. late childhood

______ **14**. Which stage of life lasts the longest?
a. infancy
b. childhood
c. adolescence
d. adulthood

15. List the three stages of childhood and give two characteristics of each stage.

____

____

____

____

____

16. Give one example of a hormone and tell what that hormone does.

17. Describe three conditions associated with aging.

CRITICAL THINKING

18. Mark's baby brother, Joey, was just born. Summarize the development he will experience during the next year.

19. Many people who have lost a loved one remain in the despair stage of grief. Explain why remaining in this stage of grief is unhealthy.

INTERPRETING GRAPHICS

Examine the diagram below, and answer the questions that follow.

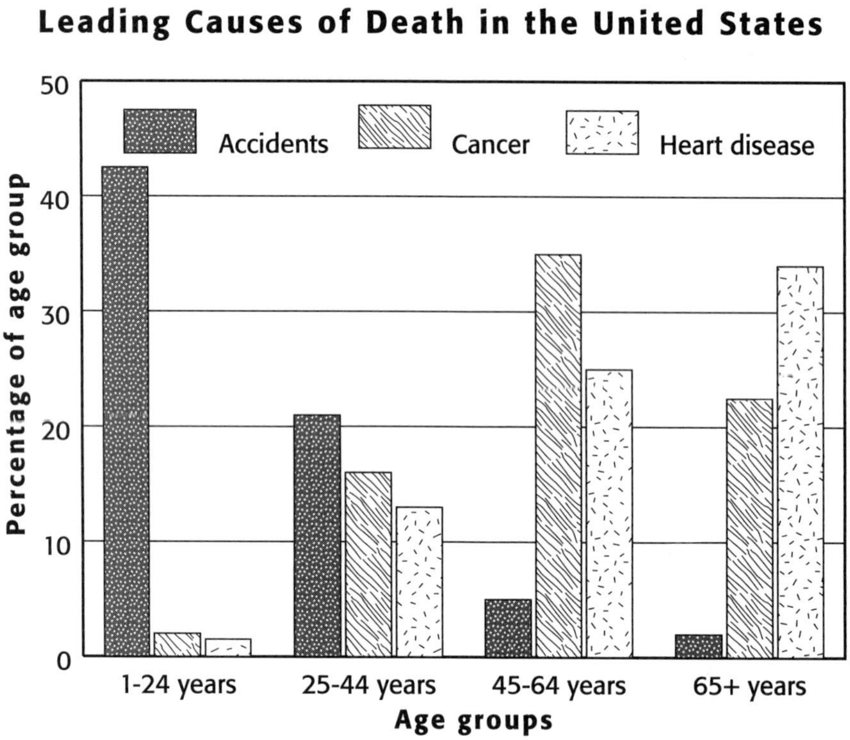

20. What is the leading cause of death for 25 to 44 year olds?

____

____

21. Which cause of death steadily declines with age?

____

____

22. Which cause of death steadily increases with age?

____

____

Assessment

Performance-Based Assessment

Stages of Human Development

INTRODUCTION

You've read about human growth and development. Now you will have a chance illustrate the process of human development.

OBJECTIVE

- Keep in mind that your teacher will be observing and grading your in-class behavior as well as your written responses. In particular, your teacher will be noting your ability to follow the given procedures, how well you follow classroom safety guidelines, and your methods and reasoning in solving problems.
- Try not to let what others are doing influence your work. Remember that a problem often has several acceptable solutions.
- Do not talk to other students unless you are working in a group. Talk only to members of your group and try not to disturb other students.
- Use only the materials provided.

SAFETY CAUTION

Use caution when using scissors.

MATERIALS AND EQUIPMENT

- markers
- scissors
- poster board

PROCEDURE

1. Make a poster illustrating the major stages of human development and the events associated with each stage.
2. Draw a picture for each stage of human development. The picture should represent the development or abilities of someone in that stage of development. The pictures can be realistic or symbolic.
3. Under each picture, describe what happens during that stage of development.
4. Hang your poster on the wall and examine the posters of your classmates.

ANALYSIS

Answer the following questions in the space provided. Support your answers by explaining your reasoning.

1. Why is it important for a mother to have healthy habits during pregnancy?

____

____

____

____

2. In what ways did other students illustrate the stages of life?

____

____

____

____

____

3. How would your poster be different if you were asked to organize the space to show the amount of time humans spend in each stage of development?

____

____

____

____

____

4. Does your poster represent mainly physical, mental, emotional, or social development? Why do you think this is?

____

____

____

____

____

Name ______________________ Class ______________ Date ____________

Activity

Datasheet for In-Text Activity

How You've Changed

1. Collect three full-length photos of yourself—one from each stage of childhood.
2. Measure the length of your whole body and the length of your head in each photo.
3. Calculate the ratio of head length to body length for each photo by dividing the head length by the body length.

ANALYSIS

1. Make a graph of the ratio of head length to body length over time.

2. How did the ratio of your head length to your body length change as you grew?

____

____

____

Name ______________________ Class ____________ Date ____________

Activity

Life Skills: Evaluating Media Messages

Lesson: Adolescence

ADVERTISING FOR ADOLESCENTS

Watch TV commercials and look at magazine advertisements. Pick out advertisements that target adolescents. Identify the types of products that are marketed to adolescents and the techniques advertisers use to reach the adolescent audience.

1. What types of products are advertised to adolescents?

____

____

____

2. Are more advertisements targeted at boys or at girls? Explain your answer.

____

____

____

3. What techniques do advertisers use to get the attention of adolescents?

____

____

____

4. Which advertising techniques do you think are most effective? Why?

____

____

____

____

5. Pick the most effective advertisement you found and describe it below. What makes it so effective?

____

____

____

____

Name ______________________ Class ______________ Date ____________

Activity

Life Skills: Coping

Lesson: Adulthood, Aging, and Death

COPING WITH GRIEF

Your friend Kelly lives far away from you. Her grandmother just passed away. Imagine that you received the letter below from Kelly. Write a letter in return that gives Kelly suggestions for coping with her grief.

Dear Friend,

How are you? I'm not doing very well right now. Grandma just died last week. I still can't believe that she's gone. It just doesn't feel real. The funeral was yesterday and it was so sad. I don't know what I'll do without Grandma. I loved her so much.

I hope everything is OK with you.

Your friend,

Kelly

Name ______________________ Class ______________ Date ____________

Activity

Enrichment Activity

Lesson: Human Reproduction

Interview a doctor or nurse about practicing good hygiene. Find out what habits are important for good hygiene. Then, find out the consequences of poor hygiene. List your findings on a separate sheet of paper.

Now, use the information from the interview to create a pamphlet on personal hygiene. Your pamphlet should include tips on practicing good hygiene as well as the possible consequences of poor hygiene.

Lesson: Before You Were Born

Go to the library and research the stages of fetal development. Find out what occurs at each stage. Using your research and the information in your textbook, write a 5-paragraph research report on fetal development. Your report should include an introduction, a conclusion, and three body paragraphs. Outline your report before you write it.

Lesson: Infancy and Childhood

Make a poster that illustrates the three stages of childhood development. You should draw or find a picture for each stage. Then, write the developments that occur during that stage under each picture. You may want to do library research on childhood development to add to your poster. Sketch your ideas on a separate sheet of paper.

Lesson: Adolescence

Go to the library and research the term *rites of passage.* Then, pick another world culture (such as the Masai tribes of East Africa) for more research. Prepare an oral presentation for the class describing the rituals and ceremonies that mark the passage of boys and girls from childhood to adulthood. Your presentation should cover the following:

- Definition of "rites of passage"
- Changing role expectations as boys become men and girls become women
- Ceremonies and rituals that mark the passage from childhood to adulthood; these may be different for boys and girls
- How changing expectations and rites of passage in the culture you studied are the same or different from those in your own culture.

Use a separate sheet to outline your presentation.

Lesson: Adulthood, Aging, and Death

Choose a disease that is associated with old age. You can choose one of the diseases listed in the textbook or one that is not. Go to the library and research the disease. Find out the following:

- What causes the disease?
- What are the symptoms of the disease?
- Is the disease treatable?
- Is the disease curable?
- How can you prevent the disease?

Using your research, produce an informational video on the disease. Your video should explain the causes, symptoms, treatments and cures, and preventive measures.

Name ______________________ Class ______________ Date ____________

Activity

Health Inventory

Growth and Development

Read each of the following statements. Think about how well it describes your behavior. Then write *always, sometimes,* or *never* in the space next to each statement.

______________________ **1.** I go along with whatever my friends are doing.

______________________ **2.** I will do something to fit in, even though I know it is dangerous.

______________________ **3.** If a friend offers me alcohol, I will accept it just to fit in.

______________________ **4.** If someone tells me smoking is cool, I will try smoking.

______________________ **5.** It is difficult for me to say no.

______________________ **6.** If a friend were to offer me drugs, I would accept them just to fit in.

______________________ **7.** It is difficult for me to walk away from unhealthy situations.

______________________ **8.** I'm afraid someone will make fun of me if I don't do the things my friends do, even if they are unhealthy.

______________________ **9.** I find myself doing things I don't want to do just because everyone else is doing them.

Score yourself on this quiz. Give yourself 0 points for each *always,* 3 points for each *sometimes,* and 5 points for each *never.* Write your score here ______.

Less than 25 Excellent—You are on top of your health and in control of your life.

25–29 Good—You recognize the need to care for yourself and often do so.

35–39 Fair—You are working on some areas of health, but you need to focus on improving your overall life skills.

40–45 You need to focus on your health and wellness. Examine the nine life skills and choose a starting point.

Name ______________________ Class ______________ Date ____________

Activity

Health Behavior Contract

Growth and Development

My Goals: I, ________________________________, will accomplish one or more of the following goals:

I will be more informed about my reproductive health.

I will talk with my parent or guardian about reproductive health.

I will perform monthly physical self-examinations and schedule regular medical checkups.

Other: ____

____

My Reasons: By improving my awareness of my reproductive health, I will be able to make good choices, such as abstaining from sexual activity. I will be able to take better care of my body by performing self-examinations and having regular medical checkups.

Other: ____

____

My Values: Personal values that will help me meet my goals are

____

____

____

My Plan: The actions I will take to meet my goals are

____

____

____

Evaluation: I will use my Health Journal to keep a log of actions I took to fulfill this contract. After 1 month, I will evaluate my goals. I will adjust my plan if my goals are not being met. If my goals are being met, I will consider setting additional goals.

Signed ______________________

Date ______________________

Name ______________________ Class ______________ Date ____________

Activity

At-Home Activity

Seeking Independence

During adolescence, you will likely seek more independence from your parents. Sit down with a parent to discuss guidelines for gaining independence.

1. What types of independence are you seeking?

____

____

____

____

2. What are the conditions for your gaining this independence?

____

____

____

____

3. What freedoms are your parents willing to grant you?

____

____

____

____

4. What are the consequences if you do not hold up your end of the deal?

____

____

____

____

The signatures below verify that our discussion has take place.

____ ______________

Student Signature Class Period

____ ______________

Parent or Guardian Signature Date

Name ______________________ Class ______________ Date ____________

Activity

Actividad En Casa

La búsqueda de la independencia personal

Durante la adolescencia, Ud. querrá más independencia de sus padres o su tutor. Discuta las normas para ganar más independencia con su padre/madre/tutor. Conteste las preguntas siguientes.

1. ¿Qué clases de independencia busca Ud.?

____

____

____

2. ¿Cuáles son las condiciones con las cuales hay que cumplir para ganar esta independencia?

____

____

____

____

3. ¿Qué privilegios está dispuesto/a su padre/madre/tutor de concederle a Ud.?

____

____

____

____

4. ¿Cuáles son las consecuencias para Ud. si no cumple con sus deberes?

____

____

____

____

Las firmas debajo de verifica que nuestra discusión tiene sucede.

______________________________ ______________________

Firma de Estudiante — Período de Clase(la Salud)

______________________________ ______________________

Firma de Padre/Madre/Tutor — Fecha

Lesson Plan

Lesson: Human Reproduction

Pacing

45 minutes

Objectives

1. Describe the structure and function of the male and female reproductive systems.
2. Identify four ways to protect your reproductive health.

Standards Covered

1.1 (partial)Explain the relationship between positive health behaviors and and the prevention of injury, illness, disease and premature death.

1.7 Explain how appropriate health care can prevent premature death and disability

7.2 Express information and opinions about health issues.

KEY	
SE = Student Edition	**ATE** = Annotated Teacher Edition
CRF = Chapter Resource File	

CHAPTER OPENER

- ❑ **Health IQ, SE** To assess student knowledge about growth and development, have students answer the Health IQ questions. Answers are at the bottom of the test.

FOCUS

- ❑ **Bellringer, ATE** Students explain why the male and female reproductive systems are different.
- ❑ **Bellringer Transparency** Use this transparency as students enter the classroom and find their seats.
- ❑ **Start Off Write, SE** Ask students to write an answer to the following question: "What are the functions of the male and female reproductive systems?"

MOTIVATE

- ❑ **Life Skill Builder, Communicating Effectively, ATE** Students make a plan to discuss reproductive health questions with a parent or doctor. **[General]**

TEACH

- ❑ **Teaching Transparency The Male Reproductive System** Use this graphic to help students understand the male reproductive system.

- ❑ **Teaching Transparency The Female Reproductive System** Use this graphic to help students understand the female reproductive system.
- ❑ **Teaching Transparency The Menstrual Cycle** Use this graphic to help students understand the menstrual cycle.
- ❑ **Activity, Learning about UTIs, ATE** This activity teaches students about urinary tract infections. **[ADVANCED]**
- ❑ **Life Skill Builder, Practicing Wellness, ATE** This activity helps students understand how to protect their reproductive health. **[GENERAL]**

CLOSE

- ❑ **Lesson Quiz, ATE** Students answer 3 questions about growth and development. **[GENERAL]**
- ❑ **Lesson Quiz, CRF** Students answer 8 questions about growth and development. **[GENERAL]**
- ❑ **Concept Review, CRF** This exercise reinforces the material covered in the lesson.**[GENERAL]**

HOMEWORK

- ❑ **Lesson Review, SE** Assign questions 1–6 for review, homework, or quiz.
- ❑ **Refusal Skills, CRF** As advice columnists, students answer a pregnant woman's question about refusing alcohol at a party. **[GENERAL]**
- ❑ **Decision-Making Skills, CRF** Students apply decision-making steps to determine if they should go to the doctor for abdominal pain that is getting worse. **[GENERAL]**
- ❑ **Concept Mapping, CRF** Students create a concept map explaining how sex cells are made. **[GENERAL]**

OTHER RESOURCE OPTIONS

- ❑ **Internet Connect** Reproductive System, HealthLinks Code HD4081. Students research Internet sources about the reproductive system.
- ❑ **go.hrw.com** For worksheets, videos, and other teaching aids related to this chapter, visit the HRW Web site and type in the keyword HD4GD6.
- ❑ **Video Select** Videos related to the chapter topics may be found at go.hrw.com. Type in the keyword HD4GD6V.
- ❑ **Guided Audio CD Program** Growth and Development The audio program is a reading of the chapter content for ELL students, auditory learners, and struggling readers.
- ❑ **Enrichment Activity, CRF** Students interview a doctor or nurse about good hygiene and the consequences of poor hygiene. **[GENERAL]**
- ❑ **Directed Reading, CRF** This worksheet guides students through the lesson content. **[BASIC]**

Lesson Plan

Lesson: Before You Were Born

Pacing

25 minutes

Objectives

1. Summarize the growth and development of humans before birth.
2. Describe three factors that affect the health of both the mother and fetus.

Standards Covered

1.1 (partial) Explain the relationship between positive health behaviors and and the prevention of injury, illness, disease and premature death.

7.2 Express information and opinions about health issues.

KEY	
SE = Student Edition	**ATE** = Annotated Teacher Edition
CRF = Chapter Resource File	

FOCUS

- ❑ **Bellringer, ATE** Students write a paragraph that describes what they think it is like inside the uterus as a fetus develops.
- ❑ **Bellringer Transparency** Use this transparency as students enter the classroom and find their seats.
- ❑ **Start Off Write, SE** Ask students to write an answer to the following question: "How long is the average pregnancy?"

MOTIVATE

- ❑ **Group Activity, Skit, ATE** Students rehearse talking to a doctor about pregnancy. **[General]**

TEACH

- ❑ **Life Skills Activity, Practicing Wellness, SE** This activity asks students to research the special dietary needs of a pregnant woman. **[General]**
- ❑ **Inclusion Strategies, ATE** This activity asks students to research human gestation. **[Advanced]**
- ❑ **Activity, Poster Project, ATE** This activity helps students understand Fetal Alcohol Syndrome. **[General]**
- ❑ **Cross-Disciplinary: Art, CRF** This activity asks students to make a poster illustrating the stages of fetal development. **[General]**

CLOSE

❑ **Lesson Quiz, ATE** Students answer 3 questions about fetal development. **[GENERAL]**

❑ **Lesson Quiz, CRF** Students answer 5 questions about fetal development. **[GENERAL]**

❑ **Concept Review, CRF** This exercise reinforces the material covered in the lesson. **[GENERAL]**

HOMEWORK

❑ **Lesson Review, SE** Assign questions 1–5 for review, homework, or quiz.

❑ **Directed Reading, CRF** This worksheet guides students through the lesson content. **[BASIC]**

OTHER RESOURCE OPTIONS

❑ **Internet Connect** Before Birth, HealthLinks Code HD4013. Students research Internet sources about fetal development.

❑ **go.hrw.com** For worksheets, videos, and other teaching aids related to this chapter, visit the HRW Web site and type in the keyword HD4GD6.

❑ **Video Select** Videos related to the chapter topics may be found at go.hrw.com. Type in the keyword HD4GD6V.

❑ **Guided Audio CD Program** Growth and Development. The audio program is a reading of the chapter content for ELL students, auditory learners, and struggling readers.

❑ **Enrichment Activity, CRF** Students research and write a report on the stages of fetal development. **[ADVANCED]**

Lesson Plan

Lesson: Infancy and Childhood

Pacing

20 minutes

Objectives

1. Summarize development during infancy and childhood.

Standards Covered

None

KEY	
SE = Student Edition	**ATE** = Annotated Teacher Edition
CRF = Chapter Resource File	

FOCUS

- ❑ **Bellringer, ATE** This activity asks students to describe the characteristics of childhood development.
- ❑ **Bellringer Transparency** Use this transparency as students enter the classroom and find their seats.
- ❑ **Start Off Write, SE** Ask students to write an answer to the following question: "How do babies change as they grow into children?"

MOTIVATE

- ❑ **Group Activity, Childproof Home, ATE** Students explore how to childproof a home. **[General]**

TEACH

- ❑ **Biology Connection, Animal Babies, ATE** Students explore the development of animal babies. **[Advanced]**
- ❑ **Reading Skill Builder, Anticipation Guide, ATE** This activity asks students to predict the stages of childhood development. **[Basic]**
- ❑ **Cross-Disciplinary: Language Arts, CRF** This activity asks students to write a narrative about a significant event in their childhood. **[General]**

CLOSE

- ❑ **Lesson Quiz, ATE** Students answer 4 questions about infancy and childhood. **[General]**
- ❑ **Lesson Quiz, CRF** Students answer 5 questions about infancy and childhood. **[General]**

❑ **Concept Review, CRF** This exercise reinforces the material covered in the lesson. **[General]**

HOMEWORK

❑ **Lesson Review, SE** Assign questions 1–3 for review, homework, or quiz.

❑ **Directed Reading, CRF** This worksheet guides students through the lesson content. **[Basic]**

OTHER RESOURCE OPTIONS

❑ **Internet Connect** Human Development, HealthLinks Code HD4057. Students research Internet sources about human development.

❑ **go.hrw.com** For worksheets, videos, and other teaching aides related to this chapter, visit the HRW Web site and type in the keyword HD4GD6.

❑ **Video Select** Videos related to the chapter topics may be found at go.hrw.com. Type in the keyword HD4GD6V.

❑ **Guided Audio CD Program** Growth and Development. The audio program is a reading of the chapter content for ELL students, auditory learners, and struggling readers.

❑ **Enrichment Activity, CRF** Students make a poster that illustrates and describes the three stages of childhood development. **[Advanced]**

Lesson Plan

Lesson: Adolescence

Pacing

45 minutes

Objectives

1. Explain how hormones affect growth and development.
2. Describe physical development during adolescence.
3. Summarize the emotional physical, mental, and social changes that happen during adolescence.

Standards Covered

1.1 (partial) Explain the relationship between positive health behaviors and the prevention of injury, illness, disease and premature death.

6.1 (partial) Demonstrate the ability to apply a decision making process to health issues and problems individually and collaboratively.

KEY
SE = Student Edition **ATE** = Annotated Teacher Edition
CRF = Chapter Resource File

FOCUS

- ❑ **Bellringer, ATE** Students write a paragraph about a major physical, mental, emotional, or social change they have recently experienced.
- ❑ **Bellringer Transparency** Use this transparency as students enter the classroom and find their seats.
- ❑ **Start Off Write, SE** Ask students to write an answer to the following question: "What are some changes that happen during adolescence?"

MOTIVATE

- ❑ **Demonstration, Finding Meaning, ATE** Students define "growing up." **[GENERAL]**

TEACH

- ❑ **Teaching Transparency, Physical Changes in Boys During Puberty** Use this graphic to help students understand the physical changes in boys during puberty.
- ❑ **Teaching Transparency, Physical Changes in Girls During Puberty** Use this graphic to help students understand the physical changes in girls during puberty.

- ❑ **Using the Figure, Identifying Changes, ATE** This activity asks students to explore the physical changes in boys and girls during puberty. **[BASIC]**
- ❑ **Life Skill Builder, Using Refusal Skills, ATE** This activity asks students to write a skit about refusing to participate in unhealthy behaviors. **[GENERAL]**
- ❑ **Life Skills: Evaluating Media Messages, CRF** Students review TV and magazine ads and identify techniques advertisers use to influence audiences. **[GENERAL]**

CLOSE

- ❑ **Lesson Quiz, ATE** Students answer 3 questions about adolescence. **[GENERAL]**
- ❑ **Lesson Quiz, CRF** Students answer 5 questions about adolescence. **[GENERAL]**
- ❑ **Concept Review, CRF** This exercise reinforces the material covered in the lesson.**[GENERAL]**

HOMEWORK

- ❑ **Lesson Review, SE** Assign questions 1–4 for review, homework, or quiz.
- ❑ **Decision-Making Skills, CRF** Students apply decision-making skills when offered beer at a party. **[GENERAL]**
- ❑ **Refusal Skills, CRF** This worksheet asks students to practice refusing cigarettes. **[GENERAL]**

OTHER RESOURCE OPTIONS

- ❑ **Internet Connect** Puberty, HealthLinks Code HD4078. Students research Internet sources about puberty.
- ❑ **go.hrw.com** For worksheets, videos, and other teaching aides related to this chapter, visit the HRW Web site and type in the keyword HD4GD6.
- ❑ **Video Select** Videos related to the chapter topics may be found at go.hrw.com. Type in the keyword HD4GD6V.
- ❑ **Guided Audio CD Program** Growth and Development. The audio program is a reading of the chapter content for ELL students, auditory learners, and struggling readers.
- ❑ **Enrichment Activity, CRF** Students research and prepare an oral presentation on rites of passage in another culture. **[ADVANCED]**
- ❑ **Directed Reading, CRF** This worksheet guides students through the lesson content. **[BASIC]**

Lesson Plan

Lesson: Adulthood, Aging and Death

Pacing

45 minutes

Objectives

1. Describe development during adulthood.
2. Explain how technology has affected life expectancy in the United States.
3. List the stages of grief.

Standards Covered

1.1 (partial)Explain the relationship between positive health behaviors and and the prevention of injury, illness, disease and premature death.

4.3 Analyze the influence of technology on personal health.

7.2 Express information and opinions about health issues.

> **KEY**
> **SE** = Student Edition   **ATE** = Annotated Teacher Edition
> **CRF** = Chapter Resource File

FOCUS

- ❑ **Bellringer, ATE** Students list the responsibilities of an adult.
- ❑ **Bellringer Transparency** Use this transparency as students enter the classroom and find their seats.
- ❑ **Start Off Write, SE** Ask students to write an answer to the following question: "How do people change after they reach adulthood?"

MOTIVATE

- ❑ **Activity, Skit, ATE** Students perform a skit about becoming an adult. **[General]**

TEACH

- ❑ **Group Activity, Aging Gracefully, ATE** This activity asks students to explore the effects of exercise on aging. **[General]**
- ❑ **Debate, Older and Older, ATE** This activity asks students to debate the effects of longer life spans on society. **[Advanced]**
- ❑ **Activity, Role-Playing, ATE** This activity asks students to demonstrate the five stages of grief. **[General]**
- ❑ **Inclusion Strategies, ATE** This activity asks students to practice making appropriate comments at funerals. **[Basic]**

- ❑ **Debate, Living Longer, ATE** This activity asks students to explore why women live longer than men in the United States. **[ADVANCED]**

CLOSE

- ❑ **Lesson Quiz, ATE** Students answer 4 questions about adulthood, aging, and death. **[GENERAL]**
- ❑ **Lesson Quiz, CRF** Students answer 7 questions about adulthood, aging, and death. **[GENERAL]**
- ❑ **Concept Review, CRF** This exercise reinforces the material covered in the lesson. **[GENERAL]**

HOMEWORK

- ❑ **Lesson Review, SE** Assign questions 1–6 for review, homework, or quiz.
- ❑ **Concept Mapping, CRF** Students create a concept map about aging and adulthood. **[GENERAL]**
- ❑ **Life Skills: Coping CRF** Students write a letter to a friend whose grandmother has just died **[GENERAL]**.

OTHER RESOURCE OPTIONS

- ❑ **go.hrw.com** For worksheets, videos, and other teaching aides related to this chapter, visit the HRW Web site and type in the keyword HD4GD6.
- ❑ **Video Select** Videos related to the chapter topics may be found at go.hrw.com. Type in the keyword HD4GD6V.
- ❑ **Guided Audio CD Program** Growth and Development. The audio program is a reading of the chapter content for ELL students, auditory learners, and struggling readers.
- ❑ **Enrichment Activity, CRF** Students research and produce a video on one disease associated with aging. **[ADVANCED]**
- ❑ **Directed Reading, CRF** This worksheet guides students through the lesson content. **[BASIC]**

Lesson Plan

End of Chapter Review and Assessment

Pacing

90 minutes

KEY	
SE = Student Edition	**ATE** = Annotated Teacher Edition
CRF = Chapter Resource File	

REVIEW

- ❑ **Chapter Review, SE** Assign questions to review the material for this chapter. Use the assignment guide to customize review for lessons covered.
- ❑ **Concept Review, CRF** Vocabulary and concept review for each lesson. **[GENERAL]**

ASSESSMENT

- ❑ **Chapter Test, Growth and Development, CRF** Assign questions for general level chapter assessment. **[GENERAL]**

ALTERNATIVE ASSESSMENTS

- ❑ **Alternative Assessment, Explaining to Others, ATE** Students create a pamphlet for younger children that explains the changes that take place in the body during puberty. **[GENERAL]**
- ❑ **Performance-Based Assessment, Stages of Human Development, CRF** Students create a poster illustrating and describing the stages of human development. **[GENERAL]**
- ❑ **Test Generator, One-Stop Planner** Create a customized homework, quiz, or test using the HRW Test Generator program.
- ❑ **Test Item Listing, CRF** Use the Test Item Listing to identify questions to use in a customized homework, quiz, or test.

Parent Letter

Growth and Development

Dear Parent/Guardian,

In the next few weeks, your son or daughter's Health Education class will focus on the subject of human growth and development. Your child's understanding of human growth and development is important as he or she experiences adolescence. The chapter will explain the many changes that occur during puberty and adolescence.

Other areas covered in this chapter include human reproduction, fetal development, infancy, and childhood. We will also discuss adulthood, aging, and death.

You can actively support your child's progress in Health by communicating with him or her about the topics covered in this course. To aid this communication, I have included worksheets for you to complete with your child. Your signature at the end of this material will verify that this home interaction has taken place.

Thank you in advance for your time, cooperation, and support.

Sincerely,

____

Health Teacher

Carta a los Padres al Tutor

El crecimiento y el desarrollo del ser humano

Estimado(s) Padres/Tutor:

En las próximas semanas, la clase de Salud se enfocará en el tema del crecimiento y el desarrollo del ser humano. El entendimiento de estos dos conceptos es importante para su hijo/a durante la pubertad y la adolescencia. Este capítulo explica muchos de los cambios que ocurren durante esta etapa de la vida.

Otros asuntos tratados son la reproducción humana, el desarrollo del feto, la infancia y la niñez. También se discutirá la madurez, el envejecimiento y la muerte.

Ud. puede ayudar a su hijo/a en esta clase por hablar con él o ella sobre los temas que estudie. Para facilitar esta comunicación le mando unas hojas de trabajo para completar con su hijo/a. Su firma en la hoja verifica que Uds. discutieron esta actividad juntos.

Gracias anticipadas por su tiempo, cooperación y apoyo.

Atentamente,

____

Maestro/a de salud

Assessment

Performance-Based Assessment

Stages of Human Development

Teacher's Notes

INTRODUCTION

Students will make a poster illustrating the major stages of human development and the events associated with each stage.

TIME REQUIRED One 45-minute class period

Students will need 40 minutes to make their posters and 5 minutes to present and hang their posters.

PBA RATINGS

Easy ← 1 2 3 4 → Hard

Teacher Prep—1
Student Set-U—2
Concept Level—2
Clean Up—1

ADVANCE PREPARATION

Obtain enough markers, scissors, and poster board for each student.

SAFETY CAUTIONS

Remind students to use caution when using scissors.

PERFORMANCE

At the end of the test, students should turn in the following items:

- poster
- completed analysis questions

EVALUATION

The following is a recommended breakdown for evaluating student performance:

30% Quality and Accuracy of poster
40% Understanding of the major stages of human development
30% Answers to analysis questions

Answer Key

Directed Reading

LESSON: HUMAN REPRODUCTION

1. a parent cell that can join with another cell to create a new cell
2. to make and store sperm
3. sperm
4. c
5. bathing every day and wearing clothes that are not too tight
6. to make the female sex cell called the egg and to carry out pregnancy
7. the monthly breakdown and shedding of the lining of the uterus
8. by abstaining from sexual activity

LESSON: BEFORE YOU WERE BORN

9. fertilization
10. fetus
11. She should not use tobacco, alcohol, or other drugs. She should eat well and take vitamins. She should go to the doctor regularly.

LESSON: INFANCY AND CHILDHOOD

12. the stage of life between birth an age 1
13. a
14. Early childhood is from age 1 to age 3. Middle childhood is from age 3 to age 5. Late childhood is from age 6 to about age 11.

LESSON: ADOLESCENCE

15. Adolescence is the stage of development during which humans grow from childhood to adulthood.
16. hormone
17. sex hormones
18. c
19. Answers may vary. Sample answer: deepening voices, acne, new hair growth
20. Adolescents become interested in forming friendships, become more independent, and show more responsibility.
21. parents or other trusted adults

LESSON: ADULTHOOD, AGING, AND DEATH

22. d
23. Sample answer: raising a family; balancing family, friends, and careers
24. heredity, diet, exercise, tobacco and drug use
25. heart disease, osteoporosis, Alzheimer's disease
26. death
27. a feeling of deep sadness about a loss
28. Talk with your parents. Go to the funeral. Avoid focusing on your grief. Watch for signs of unhealthy grieving.

Concept Mapping

LESSON: HUMAN REPRODUCTION

Sex cells are made by the *male reproductive system*, which makes *sperm* in the *testes*, and the *female reproductive system* which releases *eggs* from the *ovaries*.

LESSON: ADULTHOOD, AGING, AND DEATH

Answers may vary. Sample answer: During *adulthood* people experience *aging*, which is associated with *Alzheimer's disease*, *osteoporosis*, and *heart disease*, and which leads to *death*, which is followed by *grief*.

Concept Review

1. c
2. b
3. testes
4. ovaries
5. menstruation
6. Bathe every day. Do not wear clothes that are too tight. Go to the doctor if you notice anything unusual about your body. Abstain from sex to prevent STDs.
7. One sperm and one egg come together to form a new human cell.
8. fetus
9. at 35 to 40 weeks
10. drug use, eating habits, doctor care

11. a
12. early, middle, late
13. Children start asking many questions and trying new things. They learn to speak in longer sentences, name colors, and count. They start to dress and undress themselves.
14. b
15. c
16. a
17. They cause growth spurts and cause the shape of the body to change.
18. Answers may vary. Sample answers: acne develops, body hair grows, shoulders broaden
19. Answers may vary. Sample answers: breasts develop, hips widen, menstruation begins
20. Your ability to understand complex ideas increases. You may begin to feel attracted to other people and become interested in romantic relationships. You may experience mood swings. You may feel tempted to do unhealthy or unsafe things.
21. Adulthood is the period of life that follows adolescence. Adults are expected to take responsibility for their well-being.
22. Advance in technology, such as vaccines, surgical techniques, and medications, have increased life expectancy.
23. denial, anger, bargaining, despair, acceptance

Refusal Skills

LESSON: HUMAN REPRODUCTION

Answers may vary. Sample letter: Dear Worried, Congratulations! Having a baby is certainly a cause for celebration. However, drinking alcohol can harm the fetus inside you. If someone offers you alcohol, you can simply say no. If that person pressures you further, you can offer a reason, offer an alternative, or walk away. Your husband should support your decision not to drink alcohol, and he should help you protect the health of your fetus. Sincerely, Know-It-All

LESSON: ADOLESCENCE

Answers may vary. Sample answers:

1. I would tell Michael I do not want to smoke.
2. could suggest that we play basketball instead.
3. I would tell him that smoking is dangerous and that it's illegal to smoke at our age.
4. I could go home.
5. I could choose activities that do not involve smoking.
6. I have friends who do not smoke. They will stand by my decision not to smoke.

Decision-Making Skills

LESSON: HUMAN REPRODUCTION

Answers may vary. Sample answers:

1. whether I will see a doctor
2. My health is important to me.
3. I could try to ignore the pain, or I could go to the doctor.
4. If I ignore the pain, something serious could be wrong and I could get very sick. If I go to the doctor, he or she will be able to help me.
5. I will go to the doctor.
6. I am happy I went to the doctor because now the pain is gone. I shouldn't feel embarrassed to see the doctor.

LESSON: ADOLESCENCE

Answers may vary. Sample answer: I did not want to take the beer, so I said "No, thanks." She kept insisting, but I told her I really didn't want it. She finally backed off and most of the other guests started drinking their beer. This made me really uncomfortable, so I said I had to go home, and left the party. I'm not interested in going to any more parties like that.

Cross-Disciplinary: Art

LESSON: BEFORE YOU WERE BORN

Answers may vary. Ensure that students' posters include each stage of fetal development.

Cross-Disciplinary: Language Arts

LESSON: INFANCY AND CHILDHOOD

Answers may vary, but stories should describe a significant event in the student's childhood.

Quiz

LESSON: HUMAN REPRODUCTION

1. d
2. f
3. a
4. c
5. b
6. h
7. g
8. e

LESSON: BEFORE YOU WERE BORN

1. c
2. a
3. d
4. b
5. c

LESSON: INFANCY AND CHILDHOOD

1. d
2. a
3. b
4. c
5. b

LESSON: ADOLESCENCE

1. c
2. a
3. d
4. c
5. d

LESSON: ADULTHOOD, AGING, AND DEATH

1. c
2. d
3. a
4. f
5. g
6. e
7. b

Chapter Test

1. grief
2. fertilization
3. menstruation
4. sperm
5. fetus
6. adolescence
7. b
8. a
9. c
10. d
11. d
12. c
13. b
14. d
15. Answers may vary. Sample answer: During early childhood, you learn to walk and begin to form short sentences. During middle childhood you learn to name colors and dress yourself. During late childhood you learn to read and become more interested in social relationships.
16. Answers may vary. Sample answer: The human growth hormone causes your bones and muscles to grow.
17. Heart disease damages the heart tissue and may lead to heart attacks. Osteoporosis is a condition in which bones lose a lot of their mass and strength. Alzheimer's disease leads to memory loss and thinking problems.
18. Joey's height and weight will increase rapidly. By about 3 months of age, Joey will be able hold his head up and look around, and recognize his parents. By 7 months of age, Joey will be able to roll over and will have learned to play with toys. By age 1, he will be able to crawl, stand, and cruise, and he may be able to walk and say a few words.
19. People who do not complete the grieving process cannot move on with their lives. Staying in the despair stage of grief may lead to depression.
20. accidents
21. accidents
22. heart disease

Performance-Based Assessment

See the Performance-Based Assessment Teacher's Notes.

1. Answers may vary. Sample answer: so that the baby will be healthy and the pregnancy will proceed normally
2. Answers may vary. Accept all reasonable answers.
3. Answers may vary. Accept all reasonable answers.

4. Answers may vary. Accept all reasonable answers.

Datasheet for In-Text Activity

1. Answers may vary. Accept all reasonable answers.
2. Answers may vary, but the graphs should reflect an increase in body size in relation to head size as students age.

Life Skills: Evaluating Media Messages

LESSON: ADOLESCENCE

Answers may vary. Sample answers:

1. acne medication, cosmetics, hair care, feminine hygiene products, clothing, electronics
2. There are more ads targeted at girls, such as cosmetics, hair care, and feminine hygiene products.
3. loud music, bright colors, teenage slang
4. Accept all reasonable answers.
5. Accept all reasonable answers.

Life Skills: Coping

LESSON: ADULTHOOD, AGING AND DEATH

Answers may vary, but should express sympathy and tell Kelly her grief is natural and that you will support her during her grief.

Enrichment Activities

LESSON: HUMAN REPRODUCTION

Students should create a pamphlet on personal hygiene. Pamphlets should include tips on how to practice good hygiene as well as the possible consequences of poor hygiene.

LESSON: BEFORE YOU WERE BORN

Students should write a 5-paragraph research report on fetal development. Reports should include some information that is not in the textbook.

LESSON: INFANCY AND CHILDHOOD

Students' posters should include a drawing for each stage of childhood development with a description under each.

LESSON: ADOLESCENCE

Students' presentations should cover social roles, describe rites of passage, and compare and contrast them with those they are familiar with.

LESSON: ADULTHOOD, AGING, AND DEATH

Students' videos should cover the causes, symptoms, treatments and cures, and preventive measures of their chosen disease.

Health Inventory

Student responses will vary. Students should have the option of keeping their responses private. This worksheet might be used to start a discussion on resisting peer pressure to engage in unhealthy activities.

Health Behavior Contract

Accept all reasonable responses.

At-Home Activity

Answers may vary. Sample answers:

1. later curfew, staying home alone, babysitting brothers or sisters, going on dates
2. I must act responsibly. I must let my parents know where I'm going, who I'll be with, and what time I'll be home.
3. they will give me the independence I am requesting.
4. I will lose some of my independence.

TEST ITEM LISTING

Growth and Development

MULTIPLE CHOICE

1. The process in which one sperm andone egg come together to form a new human cell is called
   a. reproduction.
   b. incubation.
   c. fertilization.
   d. menstruation.

   Answer: C Difficulty: 1 Section: 2 Objective: 1

2. At which stage in fetal development does the heart start to beat?
   a. 6 weeks
   b. 12 weeks
   c. 13 weeks
   d. 20 weeks

   Answer: A Difficulty: 1 Section: 2 Objective: 1

3. At which stage in fetal development does a fetus react to sound?
   a. 6 weeks
   b. 12 weeks
   c. 13 weeks
   d. 20 weeks

   Answer: D Difficulty: 1 Section: 2 Objective: 1

4. At which stage in fetal development are arms and legs formed?
   a. 6 weeks
   b. 12 weeks
   c. 13 weeks
   d. 20 weeks

   Answer: B Difficulty: 1 Section: 2 Objective: 1

5. At which stage in fetal development can a fetus make a fist?
   a. 6 weeks
   b. 12 weeks
   c. 13 weeks
   d. 20 weeks

   Answer: C Difficulty: 1 Section: 2 Objective: 1

6. During which stage of life do humans learn to read and write and connect ideas?
   a. infancy
   b. early childhood
   c. middle childhood
   d. late childhood

   Answer: D Difficulty: 1 Section: 3 Objective: 1

7. During which stage of life do humans learn to crawl?
   a. infancy
   b. early childhood
   c. middle childhood
   d. late childhood

   Answer: A Difficulty: 1 Section: 3 Objective: 1

8. During which stage of life do humans learn to walk and run?
   a. infancy
   b. early childhood
   c. middle childhood
   d. late childhood

   Answer: B Difficulty: 1 Section: 3 Objective: 2

9. During which stage of life do humans learn to dress and undress themselves?
   a. infancy
   b. early childhood
   c. middle childhood
   d. late childhood

   Answer: C Difficulty: 1 Section: 3 Objective: 1

10. Which of these is NOT a characteristic of infancy?
   a. rapid increase in height
   b. forming short sentences
   c. crawling
   d. rapid increase in weight

   Answer: B Difficulty: 1 Section: 3 Objective: 1

11. During which stage of human development do humans grow from childhood to adulthood?
   a. infancy
   b. late childhood
   c. adolescence
   d. None of the above

   Answer: C Difficulty: 1 Section: 4 Objective: 2

12. A chemical that is made in one part of the body and causes a change in a different part of the body is called a(n)
   a. hormone.
   b. growth spurt.
   c. ovary.
   d. gland.

   Answer: A Difficulty: 1 Section: 4 Objective: 1

13. Hormones that affect your reproductive system are called
   a. ovaries.
   b. adrenal glands.
   c. testes.
   d. sex hormones.

   Answer: D Difficulty: 1 Section: 4 Objective: 1

14. During which stage of development does the reproductive systems become mature?
   a. infancy
   b. childhood
   c. puberty
   d. adulthood

   Answer: C Difficulty: 1 Section: 4 Objective: 2

15. Which ofthe following is a characteristic of adolescence?
   a. growth spurts
   b. acne
   c. hair growth
   d. All of the above

   Answer: D Difficulty: 1 Section: 4 Objective: 2

16. Which of the following is NOT a part of the male reproductive system?
    a. uterus
    b. testes
    c. vas deferenz
    d. scrotum

    Answer: A   Difficulty: 1   Section: 1   Objective: 1

17. Which organ holds the fetus during pregnancy?
    a. vagina
    b. ovary
    c. uterus
    d. scrotum

    Answer: C   Difficulty: 1   Section: 2   Objective: 1

18. Which of the following is NOT an example of protecting your sexual health?
    a. bathing every day
    b. visiting a doctor regularly
    c. abstaining from sex
    d. wearing tight clothing

    Answer: D   Difficulty: 1   Section: 1   Objective: 2

19. At what point during fetal development does the fetus begin to move?
    a. 6 weeks
    b. 12 weeks
    c. 14 weeks
    d. 16 weeks

    Answer: D   Difficulty: 1   Section: 2   Objective: 1

20. About how long after fertilization is a baby born?
    a. 20 weeks
    b. 30 weeks
    c. 40 weeks
    d. 50 weeks

    Answer: C   Difficulty: 1   Section: 2   Objective: 1

21. During which stage of development do children learn to walk?
    a. infancy
    b. early childhood
    c. middle childhood
    d. late childhood

    Answer: B   Difficulty: 1   Section: 3   Objective: 1

22. Which stage of life lasts the longest?
    a. infancy
    b. childhood
    c. adolescence
    d. adulthood

    Answer: D   Difficulty: 1   Section: 5   Objective: 1

23. Which is the male sex cell?
    a. sperm
    b. egg
    c. ovary
    d. testis

    Answer: A   Difficulty: 1   Section: 1   Objective: 1

24. Which is the female sex cell?
    a. sperm
    b. egg
    c. ovary
    d. testis

    Answer: B   Difficulty: 1   Section: 1   Objective: 1

25 Which organs make eggs?
    a. ovaries
    b. testes
    c. vas deferens
    d. fallopian tubes

    Answer: A   Difficulty: 1   Section: 1   Objective: 1

26. What is the monthly breakdown and shedding of the uterus lining called?
    a. fertilization
    b. reproduction
    c. menstruation
    d. puberty

    Answer: C Difficulty: 1 Section: 1 Objective: 1

27. What is a developing human called?
    a. ovary
    b. egg
    c. uterus
    d. fetus

    Answer: D Difficulty: 1 Section: 2 Objective: 1

28. The time between birth and 1 year of age is called
    a. infancy.
    b. childhood.
    c. adolescence.
    d. adulthood

    Answer: A Difficulty: 1 Section: 3 Objective: 1

29. Which is a type of hormone that controls much of the growth and function of men's bodies?
    a. epinephrine
    b. thyroxine
    c. estrogen
    d. testosterone

    Answer: D Difficulty: 1 Section: 4 Objective: 1

30. Which is a type of hormone that controls much of the growth and function of women's bodies?
    a. epinephrine
    b. thyroxine
    c. estrogen
    d. testosterone

    Answer: C Difficulty: 1 Section: 4 Objective: 1

COMPLETION

31. A feeling of deep sadness about a loss is called ____________________.

    Answer: grief Difficulty: 1 Section: 5 Objective: 3

32. During ____________________, one sperm and one egg come together to form a new human cell.

    Answer: fertilization Difficulty: 1 Section: 2 Objective: 1

33. The breakdown and shedding of the uterine lining is called ____________________.

    Answer: menstruation
    Difficulty: 1 Section: 1 Objective: 1

34. The sex cells made by males are called ____________________.

    Answer: sperm Difficulty: 1 Section: 1 Objective: 1

35. A developing human inside the uterus is called a ____________________.

    Answer: fetus Difficulty: 1 Section: 2 Objective: 1

36. The stage of human development between childhood and adulthood is called ____________________.

    Answer: adolescence Difficulty: 1 Section: 4 Objective: 2

37. The ____________________ are the organs that make eggs and the hormones estrogen and progesterone.
Answer: ovaries Difficulty: 1 Section: 1 Objective: 1

38. The period from 1 year of age to age 11 is called ____________________.
Answer: childhood Difficulty: 1 Section: 3 Objective: 1

39. A(n) ____________________ is a chemical that is made in one part of the body and causes a change in a different part of the body.
Answer: hormone Difficulty: 1 Section: 4 Objective: 1

40. The stage of development when the reproductive system becomes mature is called ____________________.
Answer: puberty Difficulty: 1 Section: 4 Objective: 2

41. The period of life that lasts from age 18 until death is called ____________________.
Answer: adulthood Difficulty: 1 Section: 5 Objective: 1

42. A condition in which bones lose a lot of their mass and strength is called ____________________.
Answer: Osteoporosis
Difficulty: 1 Section: 5 Objective: 2

43. The point when all necessary life functions stop is called ____________________.
Answer: death Difficulty: 1 Section: 5 Objective: 3

SHORT ANSWER

44. List the three stages of childhood and give two characteristics of each stage.
Sample answer:
During early childhood, you learn to walk and begin to form short sentences. During middle childhood you learn to name colors and dress yourself. During late childhood you learn to read and become more interested in social relationships.
Difficulty: 2 Section: 1 Objective: 1

45. Give one example of a hormone and tell what that hormone does.
Sample answer: The human growth hormone causes your bones and muscles to grow.
Difficulty: 2 Section: 2 Objective: 1

46. Describe three conditions associated with aging.
Sample answer:
Heart disease damages the heart tissue and may lead to heart attacks. Osteoporosis is a condition in which bones lose a lot of their mass and strength. Alzheimer's disease leads to memory loss and thinking problems.
Difficulty: 2 Section: 5 Objective: 2

47. What is a sex cell?
Answer:
A sex cell is a parent cell that can join with another sex cell to create a new cell that contains all the information needed to develop into a new human being.
Difficulty: 2 Section: 1 Objective: 1

48. What are the two sex cells?
Answer: The male sex cell is sperm, and the female sex cell is the egg.
Difficulty: 2 Section: 1 Objective: 1

49. What is the best way for both men and women to prevent contracting a sexually transmitted disease?
Answer: by avoiding sexual activity
Difficulty: 2 Section: 1 Objective: 2

50. What is the length of time between fertilization and birth?
    Answer: about 9 months
    Difficulty: 1 Section: 2 Objective: 1

51. What important change occurs during the sixth week after fertilization?
    Answer: The baby's heart starts to beat.
    Difficulty: 1 Section: 1 Objective: 1

52. At what point does a developing fetus begin to move?
    Answer:
    Most fetuses begin to move by 16 weeks. The mother may not feel these movements right away.
    Difficulty: 1 Section: 2 Objective: 1

53. What span of time is called infancy?
    Answer: birth to age 1
    Difficulty: 1 Section: 3 Objective: 1

54. What is "cruising"?
    Answer: Cruising is walking by leaning on furniture or other objects for support.
    Difficulty: 1 Section: 3 Objective: 1

55. During what stage of childhood do you typically learn to talk and walk?
    Answer: early childhood
    Difficulty: 1 Section: 3 Objective: 1

56. During which stage of childhood do you typically learn to write in cursive and perform other complex tasks?
    Answer: late childhood
    Difficulty: I Section: 3 Objective: 1

57. Define *hormone*.
    Answer:
    A hormone is a chemical that is made in one part of the body and causes a change in a different part of the body.
    Difficulty: 1 Section: 4 Objective: 1

58. Name one sex hormone that controls much of the growth of the female body and one sex hormone that controls the growth of the male body.
    Answer:
    Estrogen affects the development of females. Testosterone affects the development of males.
    Difficulty: 1 Section: 4 Objective: 1

59. What is the purpose of puberty?
    Answer:
    During puberty, the reproductive system becomes mature, and people prepare for adulthood.
    Difficulty: 2 Section: 4 Objective: 1

60. Which stage of life lasts the longest? How long does it last?
    Answer: Adulthood lasts from about age 18 until death.
    Difficulty: 1 Section: 5 Objective: 1

61. What condition associated with aging often leads to memory loss?
    Answer: Alzheimer's disease
    Difficulty: 1 Section: 5 Objective: 2

62. Define *grief.*
    Answer: a feeling of deep sadness about a loss
    Difficulty: 1 Section: 5 Objective: 3

63. Describe one healthy way to deal with grief.
    Answer:
    Sample answer: Talk with parents, trusted adults, or friends about how you feel.
    Difficulty: 2 Section: 5 Objective: 3

64. Explain how sexually transmitted diseases are spread.
    Answer: by sexual activity with an infected person
    Difficulty: 1 Section: 1 Objective: 2

65. Explain how menstrual cramps can usually be treated.
    Answer:
    Cramps can usually be treated with over-the-counter pain relievers. But if they are severe or last a long time, women should see a doctor.
    Difficulty: 1 Section: 1 Objective: 2

66. Describe the developments that occur during middle childhood.
    Sample answer:
    Children start asking many questions and trying new things. They learn to speak in longer sentences, name colors, and count. They also start to dress and undress themselves.
    Difficulty: 2 Section: 3 Objective: 1

67. What are sex hormones?
    Answer: hormones that affect your reproductive system
    Difficulty: 1 Section: 4 Objective: 1

68. At what age do girls reach their full adult height? At what age do boys reach their full adult height?
    Answer:
    Girls reach their full adult height by the time they are 16 or 17 years old. Boys reach their full adult height by the time they are 18 or 19 years old.
    Difficulty: 1 Section: 4 Objective: 2

69. List the five stages of grief.
    Answer: denial, anger, bargaining, despair, acceptance
    Difficulty: 1 Section: 5 Objective: 3

ESSAY

70. Mark's baby brother, Joey, was just born. Summarize the development he will experience during the next year.
    Sample answer:
    Joey's height and weight will increase rapidly. By about 3 months of age, Joey will be able hold his head up and look around, and recognize his parents. By 7 months of age, Joey will be able to roll over and will have learned to play with toys. By age 1, he will be able to crawl, stand, and cruise, and may be able to walk and say a few words.
    Difficulty: 2 Section: 7 Objective: 1

71 Many people who have lost a loved one remain in the despair stage of grief. Explain why remaining in this stage is unhealthy.

Sample answer:

People who do not complete the grieving process cannot move on with their lives. Staying in the despair stage of grief may lead to depression.

Difficulty: 2 Section: 5 Objective: 3

INTERPRETING GRAPHICS

Examine the diagram below, and answer the questions that follow.

Leading Causes of Death in the United States

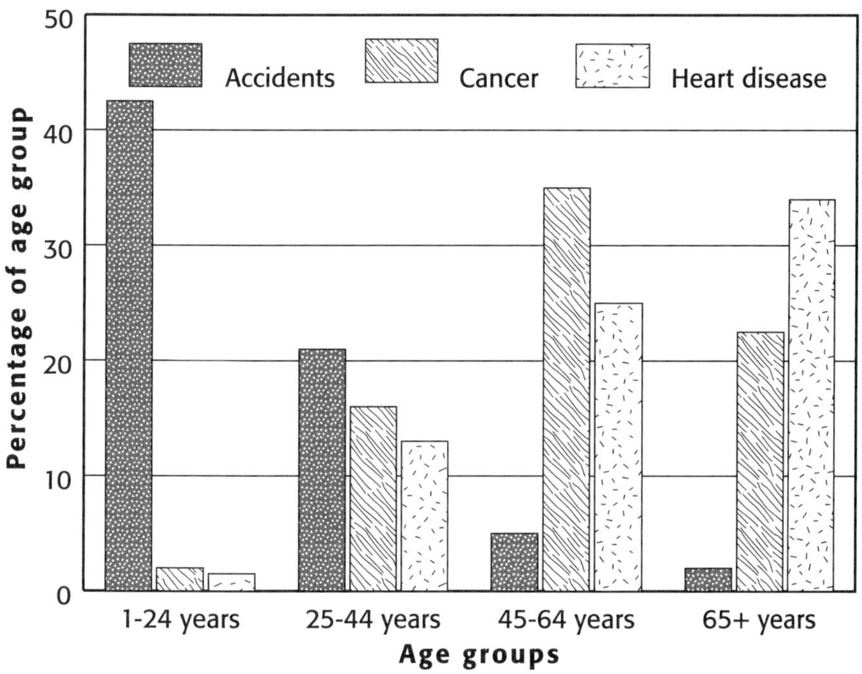

72. What is the leading cause of death for 25 to 44 year olds?
    Answer: accidents Difficulty: 1 Section: 5 Objective: 1
73. Which cause of death steadily declines with age?
    Answer: accidents Difficulty: 1 Section: 5 Objective: 2

MATCHING

a. sperm
b. testes
c. egg
d. ovaries
e. menstruation
f. sex cell
g. jock itch
h. uterus
i. adulthood
j. death
k. grief
l. aging
m. osteoporosis
n. Alzheimer's disease
o. life expectancy
p. fertilization
q. fetus
r. hormone
s. puberty

74. ____ female organs that release eggs
    Answer: D Difficulty: 1 Section: 1 Objective: 1

75. ____ parent cell that can join with another cell to create a new cell
Answer: F Difficulty: 1 Section: 1 Objective: 1
76. ____ sex cells made by males
Answer: A Difficulty: 1 Section: 1 Objective: 1
77. ____ female sex cell
Answer: C Difficulty: 1 Section: 1 Objective: 1
78. ____ organs that make sperm and testosterone
Answer: B Difficulty: 1 Section: 1 Objective: 1
79 ____ organ that holds the fetus during pregnancy
Answer: H Difficulty: 1 Section: 1 Objective: 1
80. ____ fungal infection of the groin skin
Answer: G Difficulty: 1 Section: 1 Objective: 2
81. ____ monthly breakdown and shedding of the lining of the uterus
Answer: E Difficulty: 1 Section: 1 Objective: 1
82. ____ feeling of deep sadness about a loss
Answer: K Difficulty: 1 Section: 5 Objective: 23
83. ____ change that happens as adults get older
Answer: L Difficulty: 1 Section: 5 Objective: 1
84. ____ the period of life that follows adolescence
Answer: I Difficulty: 1 Section: 5 Objective: 1
85. ____ a condition that leads to memory loss and thinking problems
Answer: N Difficulty: 1 Section: 5 Objective: 2
86. ____ expected age of death
Answer: O Difficulty: 1 Section: 5 Objective: 3
87. ____ a condition in which bones lose a lot of their mass and strength
Answer: M Difficulty: 1 Section: 5 Objective: 2
88. ____ when all necessary life functions stop
Answer: J Difficulty: 1 Section: 7 Objective: 1
89. ____ stage of development when the reproductive system becomes mature
Answer: S Difficulty: 1 Section: 2 Objective: 1
90. ____ when one sperm and one egg come together to form a new human cell
Answer: P Difficulty: 1 Section: 2 Objective: 1
91. ____ chemical that is made in one part of the body and causes a change in a different part of the body
Answer: R Difficulty: 1 Section: 4 Objective: 1
92. ____ developing human inside the uterus
Answer: Q Difficulty: 1 Section: 2 Objective: 1

9997256476 3 4 5 6